Author: Hans-Jürgen Hellberg
Cover design: Hans-Jürgen Hellberg/ Jürgen Schlüsing
Cover photo: Hans-Jürgen Hellberg

Bibliographic Information of the German National Library:
The German National Library lists this publication
in the German National Bibliography; detailed bibliographic data are
available on the Internet at dnb.dnb.de.

© 2022 Hans-Jürgen Hellberg/ Jürgen Schlüsing

Production and publishing house: BoD – Books on Demand, Norderstedt

ISBN: 9 783 755 783 183

The authors, the physicist Hans-Jürgen Hellberg and the civil engineer
Dr. Karl Jürgen Schlüsing, have repeatedly found in their lectures for
first-year students of industrial engineering that the existing mathematical
basics are not sufficient to successfully acquire the scientific
fundamentals right at the beginning of their studies. For this reason, this
booklet series for mathematics and natural sciences was created.

The booklets differ from typical textbooks, which cover complete subject
areas and are usually very comprehensive. By having each booklet
represent a single topic, the student can focus specifically on the desired
topic without having to flip through an extensive textbook or various
books. The topics in the booklets are each covered in 25 to 50 pages,
with reference to other booklets where necessary. In the case of the
natural sciences, the reference is made at the appropriate place, to the
supplementary booklets of the mathematics series. In addition, the
student will find further references in the appendix.

This system allows the student to focus, to consolidate knowledge
through short repetitions and to prepare for exams quickly and easily.

Exercise books (booklets)

I   Oscillations
II Waves
III Geometrical Optics

In preparation 2021

IV Quantum Mechanics - Atomic Physics
V Electromagnetic Fields
VI Maxwell equations

For the numerous suggestions I would like to thank
Dipl.-Ing. Leniana Ibraeva and my partner Dr. Ing. Jürgen Schlüsing.

Physical basics

I Oscillations

Introduction

To get into the physics of oscillations, the spring oscillation will serve as an example in the following.

In this case, we start with a spring that is attached at the upper end and has a weight hanging from the lower end. If we now pull on the weight, the spring begins to oscillate as soon as we release the weight. The spring thus swings perpendicular to the equilibrium position, the horizontal line (abscissa).

Another example is the pendulum.

In this example, it can be a string or a clock pendulum with a weight attached to the end. If we deflect the pendulum, it begins to swing about the vertical line (ordinate). A common description is the mathematical pendulum, which consists of a suspended thread with a mass m hanging from its lower end, neglecting the mass of the thread.

Other examples are: Oscillating circuits, electrons in an antenna or in the atomic domain and many more.

For understanding, however, the description of the spring oscillation shall suffice here.

Oscillations basic knowledge

In order to describe oscillations, we must familiarize ourselves with various terms in order to then calculate them.

Basic terms are:

Amplitude, frequency, period, harmonic oscillator without damping, harmonic oscillator with damping, forced oscillations, resonance catastrophe, superposition of oscillations (resulting phenomena).

Requirements for the calculation of oscillations is the knowledge of **trigonometric functions, frequency** and **circular motion**.

Trigonometric functions: Sine, Cosine, Tangent and Cotangent (see Mathematics Basics Exercise Book 4: 1.14.5).
Basis for the understanding and the calculation of the oscillations is first the understanding of the sine and cosine function, whereby the latter is shifted only by 90° to the sine. For this reason it is sufficient to understand the sine function at first:

$$Y = A \sin (x)$$

The **amplitude** in the above function is named factor **A**. For this purpose, the sine function $y = \sin (x)$ is taken as a basis.
If we take the unit circle with the radius as a basis, then the function has the largest value 1 and the smallest value -1 (for the unit circle, the radius is set equal to 1).
The function has the stock of values $-1 \leq y \leq 1$. The value x in the parenthesis of $\sin (x)$ is called argument.

Mnemonic 1.1 Amplitude

Multiplication by a constant factor A, yields functions that have the same periodic character (like A sin (x)), but whose maximum takes larger or smaller values).

In the figure Fig. 1.1 you find the sine function A sin (x) for the amplitude (scale freely selectable) A = 1 (see sketch below). Make clear what it means for other amplitudes like A = 2 and A = 0.5.

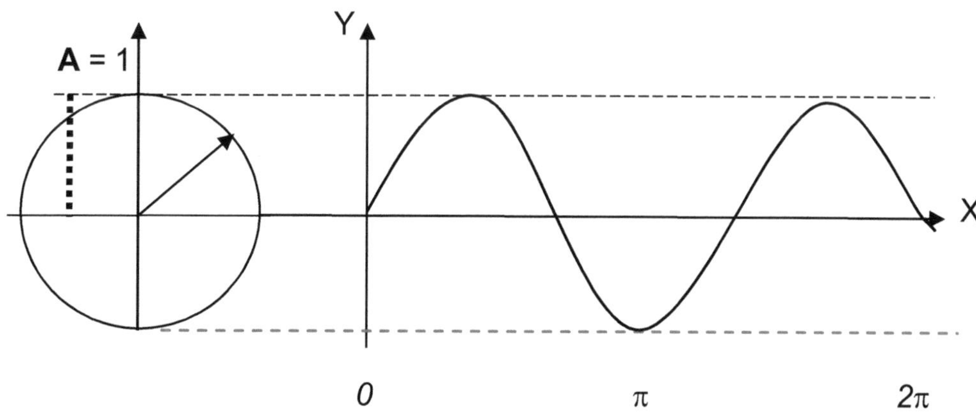

Fig. 1.1 Sine function

Meaning of frequency, angular frequency and angular velocity

Mnemonic 1.2

The **frequency** is the number of oscillations in the time interval 1 sec, where the **radian frequency** is defined as the number of oscillations in the time interval 2 / sec and the
angular velocity (also denoted by) is the angle swept in a given time.

Mnemonic 1.3

A **period** is the time required for a completed oscillation. In the case of a circular motion, a particle passes through all points of the circle in regular time intervals.

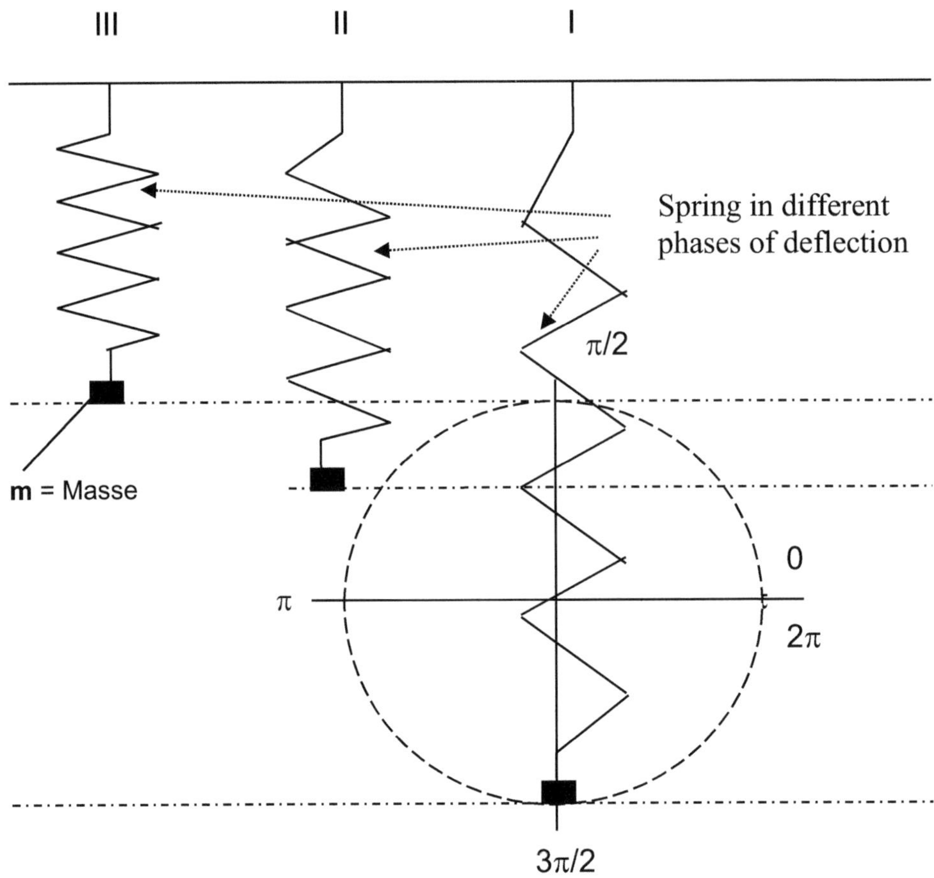

Fig. 1.2 Illustration of frequency and radian frequency

We assume a harmonic motion (Mnemonic 1.4 below)

As shown in the figure Fig. 1.2, the spring oscillates vertically. This motion can also be transferred to the circle on the right. Although the spring does not perform a circular motion, any deflection of the spring can be transferred to the circle. Rotation from right to left.

In the figure, this would be position I at the very bottom at $3\pi/2$, position II corresponds to a position between 0 and $\pi/2$ or $/\pi 2$ and π, position III corresponds to $\pi/2$ The movement starts at 0 and ends at 2π.

This circular motion thus relates then to the frequency of the revolutions per time unit of the radian frequency. In mathematical terms, a point on the circle moves from 0 to 2π.

The radian frequency is given by $\omega = 2\pi / T$. T refers the time for one complete revolution.

Mnemonic 1.4

A particle performs a **simple harmonic motion** if its displacement x relative to the coordinate origin is given as function
$Y =$ **A sin $(\omega t + \alpha)$** is given. Here $(\omega t + \alpha)$ is the phase and the original phase at time $t = 0$ with $\omega = 2\pi / T$, $\omega t + \alpha$ movement to the left and $\omega t - \alpha$ to the right.

If we consider the sine function $y = \sin(bx)$, then the argument x is multiplied by a constant factor. In the parenthesis there is a function of x: bx.

The period is the smallest number x_p for which the following applies:

$\sin(b(x + x_p))$ corresponds to $\sin(bx)$ resp.
$\sin(bx + bx_p)$ corresponds to $\sin(bx)$

since every sine function has the period 2π:
$\sin(a + 2\pi) = \sin(a)$ and thus is:

$bx = a$ and $bx_p = 2\pi$

the period of $y = \sin(bx)$ is thus:

$x_p = 2\pi / b$, so we can see that if, for example, the magnitude of b: $|b| < 1$, then the period becomes greater than 2π.

Now try to complete the table of values yourself and sketch the graph.

x	2x	sin 2x
0		
π/4		
π/2		
3π/4		
π		
5π/4		
3π/2		

Give an example to explain the meaning of the radian frequency (see above).

Determine the velocity and show that for simple harmonic motion, the acceleration of the particle is proportional and opposite to the shift.

Instruction: For this we consider the oscillation of a particle at an arbitrary point Y see fig. 1.3. The oscillation has then in dependence on the time always a certain deflection at the point Y, we can describe this at the point Y with A sin (ωt - α). If we now want to assign a velocity v to the particle, we must derive the displacement according to time (v= dy/dt) and obtain for v= ωA cos (ωt - α). Now we transfer the values to the circle on the left (below) and transfer these values to the time axis on the left (sketch below). Remember that mostly for the way the letter x is chosen, here we take y, since this is the oscillation direction, α is here =0. Values freely selectable.

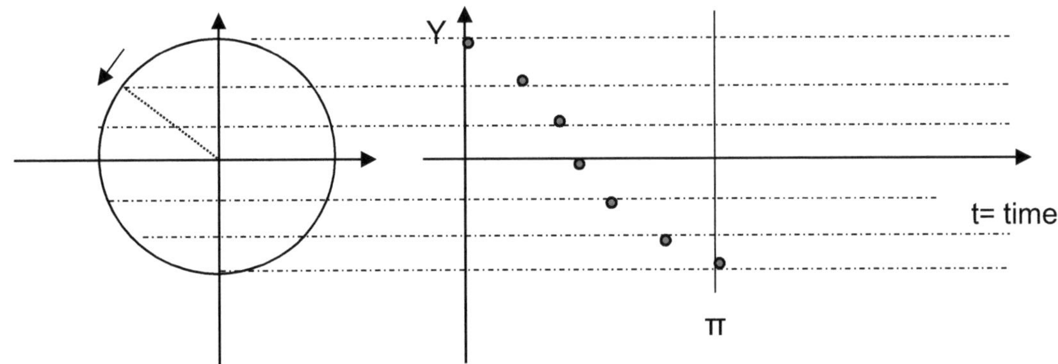

Fig.1.3 Representation of a vertical oscillation as a circular motion

Oscillatory systems,
in physics one speaks in this connection of harmonic and anharmonic oscillators:

I) The free and undamped harmonic oscillator

For our example, this means that as the deflection increases, a force acts to counteract the deflection, with the force increasing proportionally with the deflection.

Specifications:

A mass m is suspended from a spring Fig. 1.4, the spring is stretched by the distance x, the spring constant D plays a key role. It is defined as the force for deflection F/x.

Hint:

Only small deflections are to occur, then
Hooke's law is valid for the rebound force:

$F = - Dx$ with $D > 0$

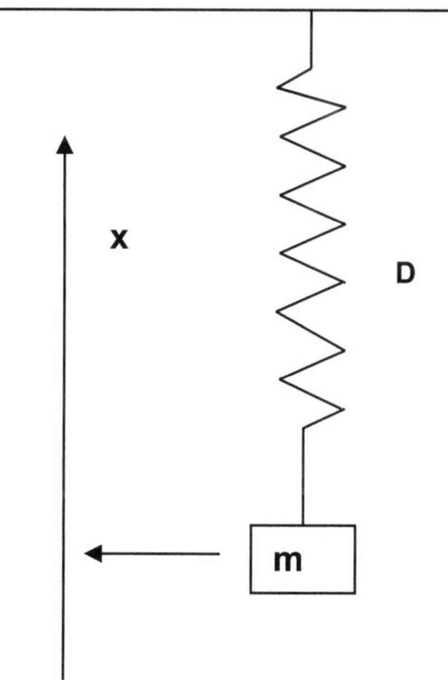

Fig. 1.4 Free and undamped harmonic oscillator (no friction takes place)

Then Newton's equation of motion is:

$$m \, \ddot{X}(t) = -Dx(t)$$, with the path X derived twice according to the time
or with the circular frequency ω:

$$\ddot{X}(t) = -\omega^2_0 \, x(t) \quad \text{(acceleration) with } \omega^2_0 = D/m.$$

But how do we arrive at this equation of motion? Since we know that
there is an oscillation, we must derive the angular velocity and get
$\omega^2_0 \, x(t)$ for the acceleration (Excursus 1.1 page 30 (radians) angular
velocity).

**Basic laws of physics are often formulated with the help of
equations** in which one finds derivatives of these equations. Of particular
importance here are the linear differential equations. The solutions are
not always very simple. Physicists use, if it is possible, the verification
principle. If one does not get further here, one tries to bring about the
solution systematically. Because the inexperienced lack the experience
for the verification principle, we will choose the second way. The
mathematics described here is a tool we cannot do without. This is also
true for the harmonic oscillator (Excursus 1.2. Part I Differential
Equations (DE) page 31 and Part II page 33).

In general, to solve a homogeneous linear differential equation of
2nd (or even 1st) order with constant coefficients, we resort to the
exponential approach. Here we assume that we have determined two
solutions:

The linear homogeneous differential solution is:

$$a_2 \, \ddot{Y} + a_1 \, \dot{Y} + a_0 \, Y = 0$$

the two different solutions are Y_1 and Y_2, then also

$$Y = C_1 Y_1 + C_2 Y_2 \quad \text{a solution of the above differential equation}$$

C_1 and C_2 are integration constants and can be any real or
complex numbers here, then we are dealing with a general
solution of the differential equation.

About the general procedure:

We again take the linear homogeneous differential equation:

$$a_2\,\ddot{Y} + a_1\,\dot{Y} + a_0\,Y = 0$$

We can solve this differential equation using the exponential approach (Mathematics Exercise Book 5: Differential Calculus):

To do this, we use the following function for Y: e^{rx}

and form from this the first and second derivative, which we again insert into the above equation and then obtain:

$$a_2 r^2 e^{rx} + a_1 r\, e^{rx} + a_0\, e^{rx} = 0$$

by excluding e^{rx}

we get: $e^{rx}\,(\,a_2\,r^2 + a_1 r + a_0\,) = 0$

e^{rx} is for every finite value of x different from zero which is why the value in the parenthesis must be zero.

We divide by e^{rx} and thus obtain a governing equation for r:

$$a_2\,r^2 + a_1\,r + a_0 = 0$$

To solve these, we resort to the 1st binomial formula (see Mathematics Exercise Book 2: 1.8):

$$(a + b)(a + b) = a^2 + 2ab + b^2$$

First we divide $a_2\,r^2 + a_1\,r + a_0 = 0$ by a_2

With this we get:

$$r^2 + \frac{a_1}{a_2}\, r + \frac{a_0}{a_2} = 0$$

Now we use the 1st binomial formula and with a = r and b = $\dfrac{a_1}{2a_2}$

We get:

$$\left(r + \frac{a_1}{2a_2}\right)\left(r + \frac{a_1}{2a_2}\right) = r^2 + 2r\frac{a_1}{a_2} + \left(\frac{a_1}{2a_2}\right)^2$$

Now we add according to our differential equation

$$r^2 + \frac{a_1}{a_2}\, r + \frac{a_0}{a_2} = 0$$

add the quadratic complement, subtract b² and obtain:

$$r^2 + 2r\frac{a_1}{2a_2} + \left(\frac{a_1}{2a_2}\right)^2 - \left(\frac{a_1}{2a_2}\right)^2 + \frac{a_0}{a_2} = 0$$

Using the 1st binomial formula (a + b)(a + b) this gives:

$$\left(r + \frac{a_1}{2a_2}\right)\left(r + \frac{a_1}{2a_2}\right) - \left(\frac{a_1}{2a_2}\right)^2 + \frac{a_0}{a_2} = 0$$

$$\left(r + \frac{a_1}{2a_2}\right)\left(r + \frac{a_1}{2a_2}\right) = \left(\frac{a_1}{2a_2}\right)^2 - \frac{a_0}{a_2}$$

Using quadratic addition, if r1 and r2 are different then we obtain with:

$$r_{1/2} = -\frac{a_1}{2a_2} \pm \sqrt{\frac{a_1^2}{4a_2^2} - \frac{a_0}{a_2}}$$

two solutions:

$$Y_1 = e^{r_1 x} \quad \text{und} \quad Y_2 = e^{r_2 x}$$

Then the general solution with the constants C_1 and C_2 is:

$$Y = C_1\, e^{r_1 x} + C_2\, e^{r_2 x}$$

It is important that the two solutions cannot be mapped for all x-values from the considered interval. These solutions must not arise from multiplication by a constant factor from each other. This kind of solutions are called linear independent. In physics, these constants are determined by the boundary conditions or constraints of the system under consideration.
The solutions themselves can differ strongly depending on the constants.

In order to get a physical result, it is now important to examine the root expression carefully, we realize that we get 3 results (the number under the root is called the radicand):

$$\left(\frac{a_1^2}{4a_2^2} - \frac{a_0}{a_2} \right)$$

Case 1) If the value is positive, then r 1 and r2 are real and different.

Case 2) If the value is negative, then r 1 and r2 are complex numbers and conjugate complex to each other.

Case 3) If the value is zero, then the double root is obtained.

$$r_1 = r_2 = -\frac{a_1}{2a_2}$$

to case 1) this one does not require any further explanation
to case 2) we have to look at this one a bit closer, because in physics we are mainly interested in real solutions, because they have a real concrete meaning.

Let's simplify and write for:

$$r_1 = a + ib \quad \text{und} \quad r_1 = a - ib$$

$$a = -\frac{a_1}{2a_2}$$

$$\text{and} \quad b = \sqrt{\frac{a_1^2}{4a_2^2} - \frac{a_0}{a_2}}$$

If necessary, please familiarize yourself with complex systems (Mathematics Exercise Book 15: Complex Numbers).

Insert $r_1 = a + ib$, $r_1 = a - ib$, and we get the general solution:

$$Y = C_1\,e^{r_1 x} + C_2\,e^{r_2 x}$$

$$Y = C_1 e^{(a+ib)x} + C_2\,e^{(a-ib)x}$$

$$Y = e^{ax}\left(C_1 e^{ibx} + C_2\,e^{-ibx}\right)$$

Let us now use Euler's equations:

$$e^{\pm ix} = \cos x \pm i \sin x$$

and replace the complex exponential function by
cos and sin functions, then we get:

$$Y = e^{ax}\left[C_1\,(\cos bx + i\,\sin\,bx) + C_2\,(\cos bx - i\,\sin bx)\right]$$

$$Y = e^{ax}\left[(C_1 + C_2\,)\cos bx + (C_1 - C_2)\,i\,\sin bx\right.$$

We now combine (C1 + C2) = A and (C1 - C2) = B to new constants and obtain with:

$$Y = e^{ax}\left[A \cos bx + i\,B \sin bx\right] \quad \text{a generally valid solution.}$$

Now we have to specify a real-valued solution function for the complex solution function solution function, it is valid:

$$a_2\,(Y_1 + i\,Y_2\,)'' + a_1\,(Y_1 + i\,Y_2)' + a_0\,(Y_1 + iY_2) = 0$$

ordered by real and imaginary values:

$$a_2\,Y''_1 + a_1 Y'_1 + a_0 Y_1 + i\,(a_2\,Y''_2 + a_1 Y'_2 + a_0 Y_2\,) = 0,$$

further applies that a complex number is zero exactly if the real part and the imaginary part are zero at the same time (in the equations Y'' $bzw.\,Y'$ stands here for once resp. twice derived).
Thus it is valid:

$$a_2 Y''_1 + a_1 Y'_1 + a_0 Y_1 = 0 \quad \text{for the real-valued part and for}$$
$$i\,(a_2 Y''_2 + a_1 Y'_2 + a_0 Y_2) \quad \text{applies, that this part = 0, if the}$$

for the real-valued part and for applies, that this part = 0, if the part rt in the parenthesis is 0.

It follows that both: Y_1 and Y_2 are solutions of the differential equation.

In our case, this means that to the complex solution:

$$Y = e^{ax}\,(A\cos bx + i\,B\sin bx)$$

also, the general real-valued solution

$$Y = e^{ax}\,(A\cos bx + B\sin bx) \quad \text{belongs.}$$

to case 3) here we get with the double root obtained above:

$$r_1 = r_2 = -\frac{a_1}{2a_2}$$

only one solution, to obtain a general one, a second one is needed. This is obtained by varying the constants, where one solution is known, namely the one we obtained by the exponential approach:

$$Y_1 = e^{r_1 x} \quad \text{and for the second solution } Y_2 = x\,e^{r_1 x}$$

and provided with the constants C_1 and C_2, we now obtain:

$$Y = C_1\,e^{r_1 x} + C_2 x\,e^{r_1 x}$$

Let us return to the free and undamped linear harmonic oscillator:

We have already established Newton's equation of motion, with acceleration:

$$\ddot{x}\,(t) = -\omega_0^2\,x(t) \quad \text{mit } \omega_0^2 = D/m$$

and thus have the differential equation:

$$\ddot{x}(t) + \omega_0^2\, x(t) = 0$$

For this we are now looking for a special solution as well as a general solution, again the exponential approach comes along:

$$X(t) = e^{rt} \quad \text{to bear and we receive:}$$

$$r^2 + \omega_0^2 = 0$$

results for $r_1 = i\,\omega_0$
results for $r_2 = -i\,\omega_0$

$$X(t) = C_1\, e^{i\omega_0 t} + C_2\, e^{-i\omega_0 t}$$

According to Euler:

$$e^{\pm i\omega_0 t} = \cos\omega_0 t \pm i\sin\omega_0 t \quad \text{eingesetzt, gilt:}$$

$$X(t) = C_1(\cos\omega_0 t + i\sin\omega_0 t) + C_2(\cos\omega_0 t - i\sin\omega_0 t)$$

$$X(t) = C_1\cos\omega_0 t + C_1\, i\sin\omega_0 t + C_2\cos\omega_0 t - C_2 i\,\sin\omega_0 t$$

$$X(t) = (C_1 + C_2)\cos\omega_0 t + (C_1 - C_2)\, i\sin\omega_0 t$$

mit A = C₁ + C₂ und B = C₁ −C₂

Thus, as we already know, our general complex solution function is:

$$X = A\cos\omega_0 t + Bi\sin\omega_0 t$$

Now we have to find a real-valued function, for this we go back to:

$$a_2\,\ddot{Y} + a_1\,\dot{Y} + a_0 Y = f(x) \ and \ f(x) = 0$$

Here the obtained solution function is a complex function Y of the real variable x.

Y = $Y_1(x)$ + i $Y_2(x)$ with the imaginary unit i, where Y_1 and Y_2 are supposed to be different from each other. Now Y_1 is the real part and Y_2 is the imaginary part of the special solutions of the differential equation. The general real-valued solution is then:

$$Y = C_1 Y_1 + C_2 Y_2$$ and in our case this is:

$$X(t) = C_1 \cos \omega_0 t + C_2 \sin \omega_0 t$$

This now allows you to define the boundary conditions:

Location at time t = 0: $\qquad\qquad X(0) = 0$

Velocity at time t = 0: $\qquad\qquad \dot{x}(0) = V_0$

Searched for: C_1 and C_2 for the special solution.

1. condition: $\quad X(0) = 0 = C_1 \cos 0 + C_2 \sin 0 = C_1$

2. results for $\dot{X}(0) = V_0 = -C_1 \omega_0 \sin 0 + C_2 \omega_0 \cos 0 = C_2 \omega_0$

$$V_0 = C_2 \omega_0$$

Converted to C_2, this results in: $C_2 = \dfrac{V_0}{\omega_0}$

The special solution is with it:

$$X(t) = \frac{V_0}{\omega_0} \sin \omega_0 t$$

And the general solution is, as we already know:

$$X(t) = C_1 \cos \omega_0 t + C_2 \sin \omega_0 t$$

This general solution is a superposition of two trigonometric functions of the same frequency. To understand this solution, it is important to look again at the addition theorems, then we find that the following is valid:

$$\cos(\varphi_1 + \varphi_2) = \cos \varphi_1 \cos \varphi_2 - \sin \varphi_1 \sin \varphi_2$$

If we now set for $C_1 = C \cos \alpha$ and for $C_2 = -C \sin \alpha$ and put this into the general solution, we get:

$$X(t) = C \cos \alpha \, \cos \omega_0 t - C \sin \alpha \sin \omega_0 \, t \text{ what we can}$$
now write as: $X(t) = C \cos (\omega_0 t + \alpha)$

About superposition: To learn a little more about amplitude, we briefly deal with superposition and determine that the two functions under consideration have the same period, although the amplitudes can be different. The sum of the two functions must again lead to a function of the same period, but it is out of phase and the amplitude depends on the amplitudes of the output functions.

In general:

$$A \sin \varphi + B \cos \varphi = C \sin(\varphi + \varphi_0)$$

with the amplitude: $\qquad C = \sqrt{A^2 + B^2}$

and the phase: $\qquad \tan \varphi_0 = \dfrac{B}{A}$

Please prove the above superposition for yourself by looking at the auxiliary sketch Fig. 1.5 (below). Help for the solution can be found in (Mathematics Exercise Book 4: 1.14.3).

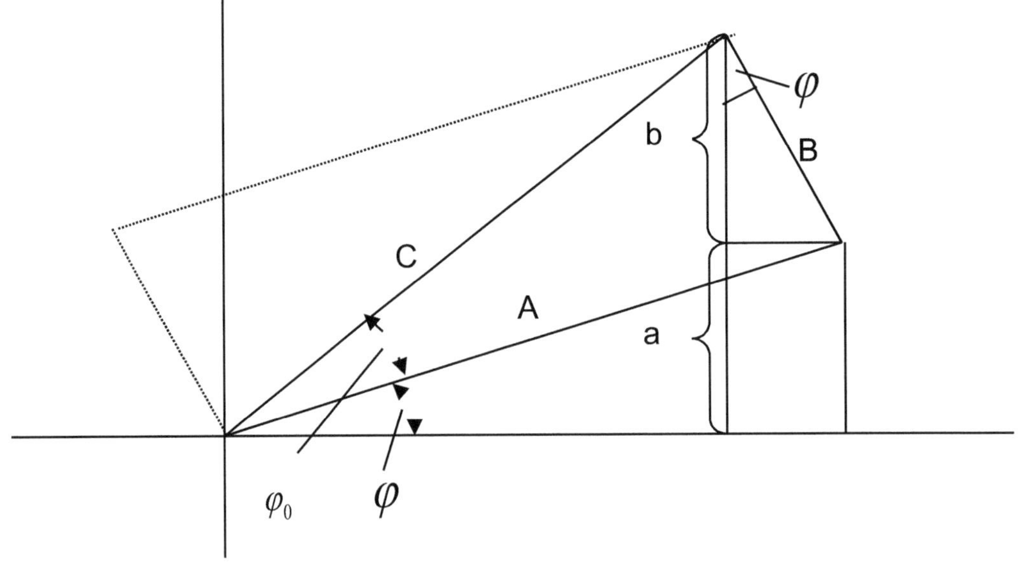

Figure. 1.5 Superposition

Having previously determined C_1 und C_2, we know that we get

$$C = \sqrt{C_1^2 + C_2^2} \quad .$$

With this we can take a look at our general solution $X(t) = C \cos(\omega_0 t + \alpha)$ again, specify the boundary conditions and then determine C und α.

As before, we choose:
Location at time t = 0: $\quad\quad X(0) = 0$
Velocity at the time t = 0: $\quad\quad \dot{x}(0) = V_0$
and receive with $\quad\quad\quad\quad\quad X(0) = 0 = C \cos(0 + \alpha)$

$$\text{for } \alpha = \frac{\pi}{2}$$

$$\dot{X}(0) = V_0 = C \omega_0 \sin\left(0 + \frac{\pi}{2}\right) = C \omega_0$$

Then the solution is with $C = \dfrac{V_0}{\omega_0}$:

$$X(t) = \frac{V_0}{\omega_0} \cos\left(\omega_0 t + \frac{\pi}{2}\right) = \frac{V_0}{\omega_0} \sin \omega_0 t$$

The special solution is thus identical to the general solution.

II) The damped harmonic linear oscillator

The harmonic oscillator represents an ideal case. In nature, motions are inhibited by frictional forces. We therefore define a friction force F_R proportional to the velocity $\dot{x}$:

$$F_r = R\,\dot{x}$$
R is the friction constant

Newton's equation of motion is thus:

$$m\,\ddot{x} = -R\,\dot{x} - Dx \quad \text{mit} \quad m\,\ddot{x} + R\,\dot{x} + Dx = 0$$

With the exponential approach $X(t) = e^{rt}$ we get the characteristic equation:

$$m\,r^2 + Rr + D = 0$$

with $\quad r_{1,2} = -\dfrac{R}{2m} \pm \sqrt{\dfrac{R^2}{4m^2} - \dfrac{D}{m}}$

depending on the radicand, we obtain two solutions:

1) $r_{1,2}$ **are real** (radicand is positive) **for:**

$$r_{1,2} = \frac{R^2}{4m^2} > \frac{D}{m}$$

then the general solution is:

$$X(t) = e^{-\frac{R}{2m}t}\left(C_1 e^{\sqrt{\frac{R^2}{4m^2} - \frac{D}{m}}\,t} + C_2 e^{-\sqrt{\frac{R^2}{4m^2} - \frac{D}{m}}\,t}\right)$$

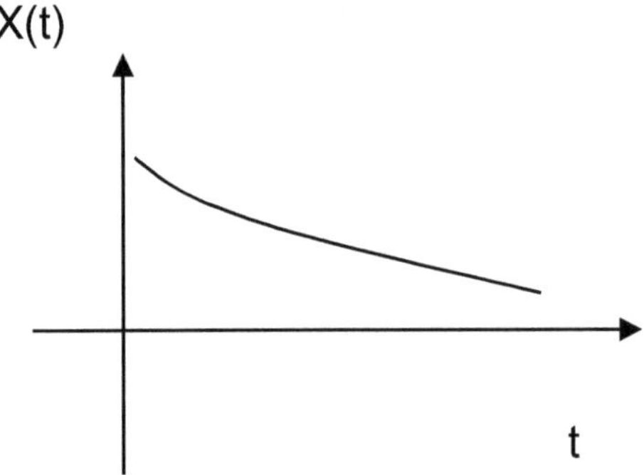

Fig. 1.6 Overall exponentially decreasing curve

We thus obtain an overall exponentially decreasing curve. In the parenthesis we have a rising and a falling function. The increasing function grows more slowly than the term in front of the parenthesis, so that applies:

$$\frac{R}{2m} > \sqrt{\frac{R^2}{4m^2} - \frac{D}{m}}$$

2) $r_{1,2}$ are conjugate complex:

$$\frac{R^2}{4m^2} < \frac{D}{m}$$

For the general solution we get:

$$X(t) = e^{-\frac{R}{2m}t}\left(C_1 \cos\sqrt{\frac{D}{m} - \frac{R^2}{4m^2}}\,t + C_2 \sin\sqrt{\frac{D}{m} - \frac{R^2}{4m^2}}\,t\right)$$

Using the addition theorems we can transform the equation as we did before, we use:

$$C \cos(x - \alpha) = C \cos x \cos \alpha + C \sin x \sin \alpha = C_1 \cos x \; C_2 \sin x$$

with $C_1 = C \cos \alpha$ and $C_2 = C \sin \alpha$ we obtain:

$$X(t) = e^{-\frac{R}{2m}t}\; C \cos\left(\sqrt{\frac{D}{m} - \frac{R^2}{4m^2}}\,t - \alpha\right)$$

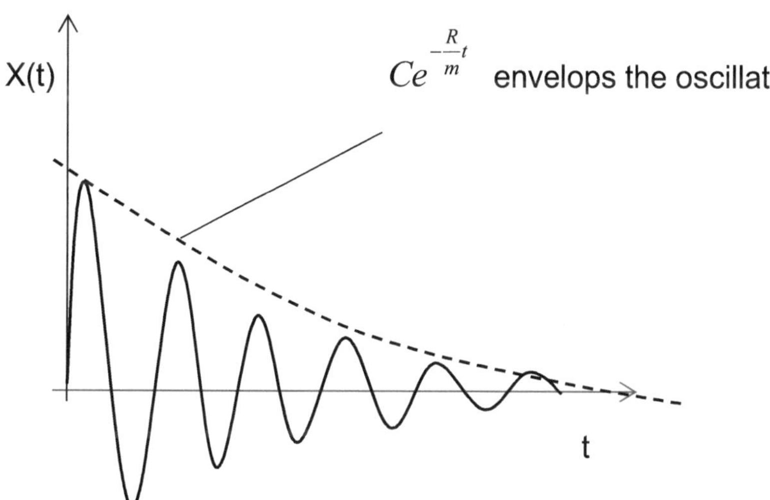

Fig. 1.7 Oscillation with exponentially decreasing amplitude

3) $r_1 = r_2 = -\dfrac{R}{2m}$

because the root expression with $\dfrac{R^2}{4m^2} = \dfrac{D}{m}$ disappears, we obtain

$$X(t) = (C_1 + C_2 x)\, e^{-\frac{R}{2m}t}$$ as solution and the course is called

aperiodic boundary case.

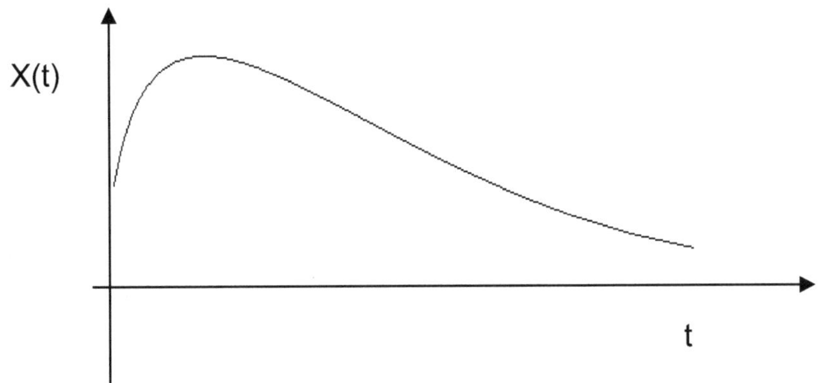

Fig. 1.8 Aperiodic limit case, here the smallest damping is described. The deflected system strives towards the equilibrium position without overshooting (i.e. there is no change of direction). The initial velocity at the start in the deflected state is 0.

III) Inhomogenous system

An example for this is the damped harmonic oscillator, on which an external force acts. One can speak here also of a forced oscillation (see mathematics exercise book 14: 5.5).

We can find this form of an oscillator in numerous areas of physics: for example, when we consider a damped harmonic oscillator and apply an external force to it.

This could be a particle which is exposed to an elastic force and on which an external oscillating motion is exerted. This is the case when a tuning fork is placed in a resonance box, forcing the walls and air in the box to vibrate. The same can be done by electromagnetic waves which are absorbed by an antenna and act on the circuit of a radio or television, thus causing forced electrical oscillations.

Let us stay with our example (Fig. 1.9 below) and let an additional external force act on the system:

$F_A = F_0 \cos(\omega_A t)$, the index A characterizes the external force
Our Newtonian equation of motion is thus:

$$m\ddot{x} + R\dot{x} + Dx = F_0 \cos(\omega_A t)$$

We have learned that the general solution of this inhomogeneous linear differential equation is equal to the sum of the general solution, the associated homogeneous differential equation plus a special solution of the inhomogeneous differential equation.

There is now the possibility to solve the equation systematically or to guess a special solution based on experience and thus make our work easier. Since the first systematic solution is very time-consuming, we try the second possibility.

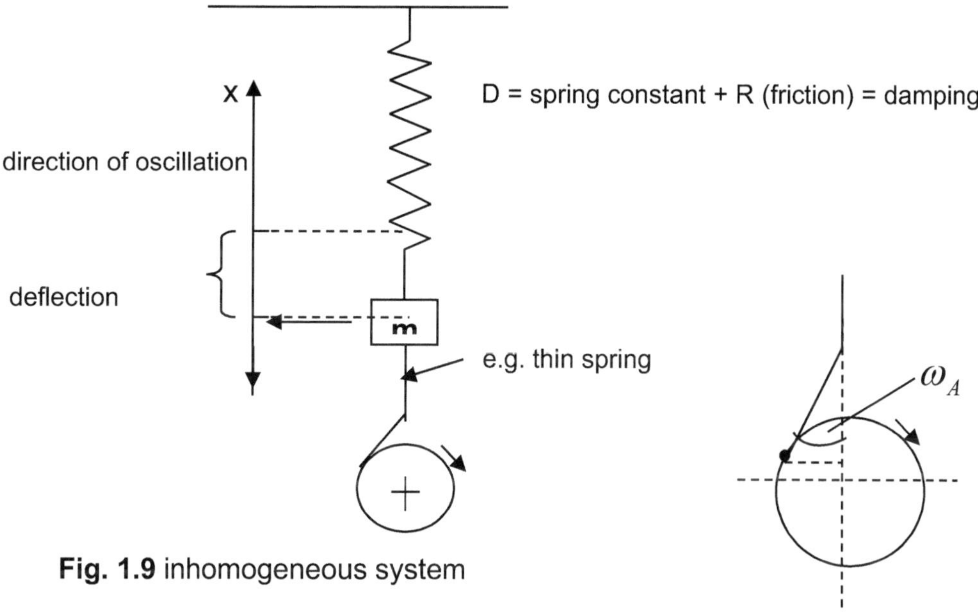

Fig. 1.9 inhomogeneous system

The form $F_A = F_0 \cos(\omega_A t)$ of the additional external force is shown in Fig. 1.9 by an eccentric at the bottom right and represents the inhomogeneous (member) of the system. The inhomogeneous member gives rise to the conjecture that the special solution X_s is of the form:

$$X_s(t) = X_0 \cos(\omega_A - \alpha)$$

But first we have to clarify what X₀ and the phase above mean.

Now it is important to know that the particle resp. the mass will not oscillate with the undamped angular frequency and also not with the damped angular frequency. On the contrary, the particle will be forced to oscillate with the angular frequency ω_A of the applied force. Finally, we consider the phase $\omega_A - \alpha$ where α is the original phase and X_0 is the amplitude. Both the amplitude and α are now fixed quantities that depend on the frequency ω_A of the exerted force.

Using the addition theorems, we obtain for:

$$X_s(t) = X_0 \cos(\omega_A t) \cos \alpha + X_0 \sin(\omega_A t) \sin \alpha$$

now we differentiate twice by t and get:

$$\dot{X}_s(t) = - X_0 \omega_A \sin(\omega_A t) \cos \alpha + X_0 \omega_A \cos(\omega_A t) \sin \alpha$$

$$\ddot{X}_s(t) = - X_0 \omega_A^2 \cos(\omega_A t) \cos \alpha - X_0 \omega_A^2 \sin(\omega_A t) \sin \alpha$$

$\ddot{X}_s(t)$ and $\dot{X}_s(t)$ is inserted into the differential equation

$$m\ddot{x} + R\dot{x} + Dx - F_0 \cos(\omega_A t) = 0:$$

$$- m(X_0 \omega_A^2 \cos(\omega_A t) \cos \alpha - X_0 \omega_A^2 \sin(\omega_A t) \sin \alpha)$$

$$+ R(- X_0 \omega_A \sin(\omega_A t) \cos \alpha + X_0 \omega_A \cos(\omega_A t) \sin \alpha) - F_0 \cos(\omega_A t) = 0$$

and the terms rearranged according to the factors of $\cos(\omega_A t)$ and $\sin(\omega_A t)$:

$$\cos \omega_A t \, (-m X_0 \omega_A^2 \cos \alpha + R X_0 \omega_A \sin \alpha + D X_0 \cos \alpha - F_0)$$

$$+ \sin \omega_A t \, (-m X_0 \omega_A^2 \sin \alpha + R X_0 \omega_A \cos \alpha + D X_0 \sin \alpha) = 0$$

If the equation is to be zero at any time, then both brackets must disappear. To do this, we recombine the contents of the brackets and obtain the two equations for X_0 and α:

First equation: $R\,\omega_A\,X_0\sin\alpha + (D - m\omega_A^2)X_0\,\cos\alpha - F_0 = 0$

resp.: $R\,\omega_A\,X_0\sin\alpha + (D - m\omega_A^2)X_0\,\cos\alpha = F_0$

Second equation: $(D - m\omega_A^2)X_0\,\sin\alpha - R\,\omega_A\,X_0\,\cos\alpha = 0$

Now follows the solution of the second equation to $X_0\cos\alpha$:

$$X_0\cos\alpha = \frac{(D - m\omega_A^2)X_0\,\sin\alpha}{R\,\omega_A}$$

Insert the result into the first equation:

$$R\,\omega_A\,X_0\sin\alpha + (D - m\omega_A^2)\,\frac{(D - m\omega_A^2)X_0\,\sin\alpha}{R\,\omega_A} - F_0 = 0$$

and solve it for $X_0\sin\alpha$ and obtain:

$$X_0\sin\alpha = \frac{R\omega_A\,F_0}{(R\,\omega_A)^2 + (D - m\,\omega_A^2)^2}$$

Accordingly, we obtain:

$$X_0\cos\alpha = \frac{F_0\,(D - m\omega_A^2)}{(R\,\omega_A)^2 + (D - m\omega_A^2)^2}$$

The division of the two equations follows:

$$\frac{X_0\,\sin\alpha}{X_0\,\cos\alpha} = \tan\alpha = \frac{\omega_A\,R}{D - m\omega_A^2}$$

Our phase is therefore $\tan\alpha = \dfrac{\omega_A\,R}{D - m\omega_A^2}$

For the amplitude the following is valid (see superposition):

At this point we recall again the superposition, where for the amplitude we have $C = \sqrt{A^2 + B^2}$.

In our case, C corresponds to X_0:

$$X_0^2 \sin^2 \alpha + X_0^2 \cos^2\alpha = X_0^2$$

We thus square and add the equations and it follows for x_0^2:

$$\frac{F_0^2 \left[\omega_A^2 R^2 + (D+m\omega_A^2)^2\right]}{\left[(D-m\omega_A^2)^2 + \omega_A^2 R^2\right]^2} = \frac{F_0^2}{(D-m\omega_A^2)^2 + \omega_A^2 R^2}$$

Thus follows for the amplitude:

$$X_0 = \frac{F_0}{\sqrt{(D - m\omega_A^2)^2 + \omega_A^2 R^2}}$$

Our general solution is then equal to:

$$X(t) = X_h(t) + X_s(t) :$$

$$X(t) = X_h + \frac{F_0}{\sqrt{(D - m\omega_A^2)^2 + \omega_A^2 R^2}} \cos(\omega_A t - \alpha)$$

$x_h(t)$ is the general solution of the homogeneous differential equation

$x_h(t)$ is an exponentially decaying function for D>0 and is practically zero after a sufficiently long time.

The mass m then only oscillates according to the following function:

$$X(t) = \frac{F_0}{\sqrt{(D-m\omega_A^2)^2 + \omega_A^2 R^2}} \cos(\omega_A t - \alpha),$$

with frequency ω_A and we speak here of an oscillation called stationary solution.

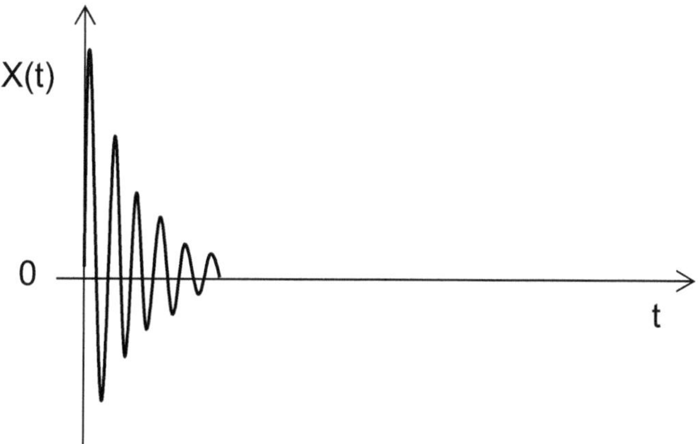

Fig. 1.10 General solution of the homogeneous differential equation Xh (t)

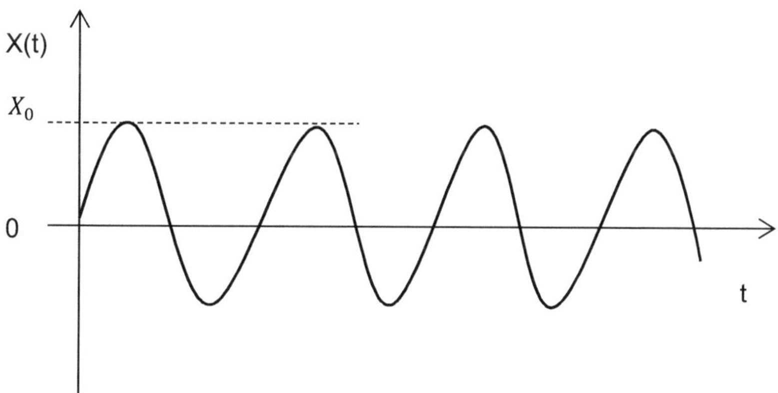

Fig. 1.11 Xs(t) represents a special solution of the differential equation

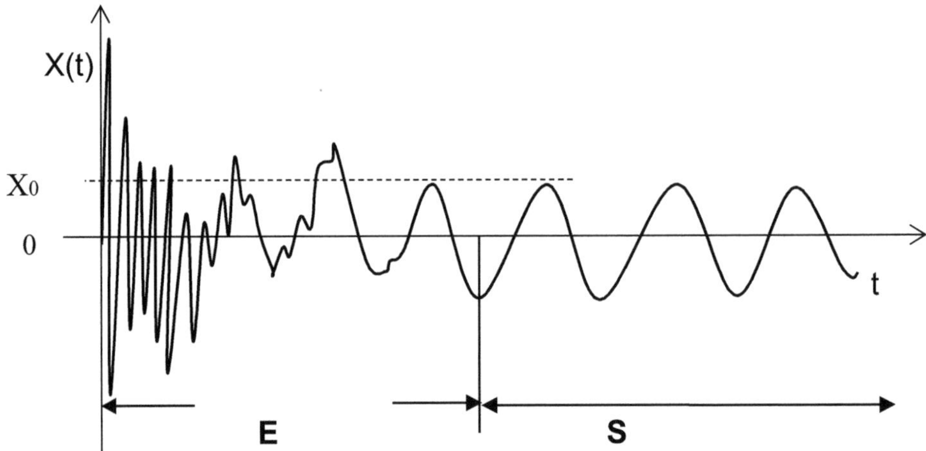

Fig. 1.10 Xh(t) + Xs(t) as a sum represents the general solution of the inhomogeneous differential equation. Here, the transient range (E) can be seen well, which then changes into a stationary state (S).

Finally, if we consider again the amplitude X0 as a function of the angular frequency ω_A of the external force acting on the mass m, then we can set the maximum value of X0 by changing ω_A:

The value for ω_A at which X0 is at maximum is the resonance frequency.

Here we are dealing with an extreme value task. The condition is:

$$\frac{dx_0}{d\omega_0} = 0$$

and we receive:

$$\omega_{AR} = \sqrt{\omega_0^2 - \frac{R^2}{2m^2}}$$

If we have R=0, then the system is undamped and the resonant frequency is equal to the oscillation frequency of the undamped oscillator. X0 is now infinitely large, since the denominator of X0 disappears. In this case, there is a resonance catastrophe.

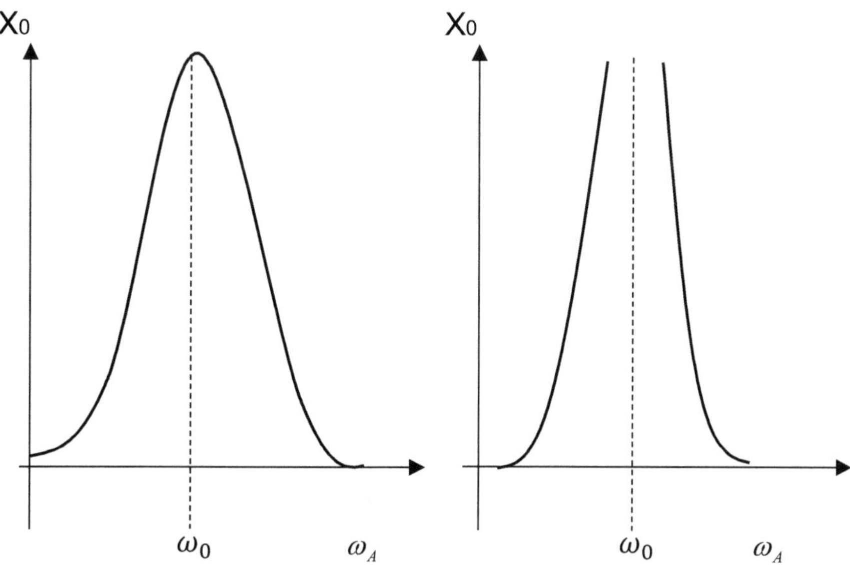

Fig. 1.12 a) Amplitude of the damped forced oscillation for $R \neq 0$,

Fig. 1.12 b) Amplitude of undamped forced oscillation for $R = 0$.

Important

The knowledge gained here, has a supporting role in many other areas of physics, which we will also see in the field of atomic and nuclear physics.

Learning Content:

You can now:
1. give examples of the term oscillations and explain their meaning
2. explain angular speed with the help of radians measure

3. say what is means harmonic oscillations.

4. explain damped harmonic oscillations,
   describe adiabatic process

5. say when we speak of a forced oscillation.

1. Mathematical supplements

Excursus 1.1 (radian measure) angular speed

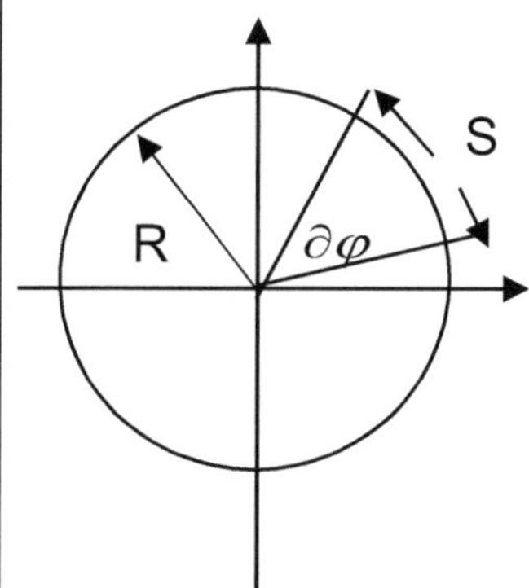

Length of the circle segment:

$$S = \partial\varphi\, R$$

Angular speed:

$$\frac{\partial\varphi}{\partial t} = \omega$$

Speed:

$$\frac{\partial S}{\partial t} = \omega R$$

How must one imagine this?

We know that circumference 2 r π, so this is U (circumference), for half the circumference we have r π = U/2, let's take an arbitrarily small angle, then we have, as shown above:

S= $\partial\varphi$ R, here we also can write $\partial U = S$

If we derive the velocity ωR again according to the time, we get the acceleration: ω^2R

We then transfer this knowledge, to the spring oscillation we are looking at.

Mathematics exercise book 4: radian measure

Excursus 1.2 Part I Differential Equations (De.)

A differential calculation is used to calculate a function with which we determine, for example, the location X as a function of time. In order to bring about the solution for our case, we repeat again the differential equations, thereby we have to do it with for us two ways:

I) Inhomogeneous linear differential equations of 2nd order with constant coefficients:

$$a_2\ddot{Y} + a_1\dot{Y} + a_0 Y = f(x) \ \text{mit} \ f(x) \neq 0$$

II) Homogeneous linear differential equation of 2nd order with constant coefficients:

$$a_2\ddot{Y} + a_1\dot{Y} + a_0 Y = f(x) \ \text{mit} \ f(x) = 0$$

Method of resolution:

to I) $\qquad\qquad a_2\ddot{Y} + a_1\dot{Y} + a_0 Y = f(x) \ \text{mit} \ f(x) \neq 0$

To solve the above differential equation, several steps are necessary. First we search Y_h (the index h stands for homogeneous) for the general solution of the homogeneous differential equation. We obtain this by setting $f(x)$ = 0

$$a_2\ddot{Y} + a_1\dot{Y} + a_0 Y = 0$$

Now we search for Y_{inh} (inh stands for inhomogeneous) of any special solution of the inhomogeneous differential equation.

We can then use the approach $Y = Y_h + Y_{inh}$ to find the general solution of the differential equation

$$a_2\ddot{Y} + a_1\dot{Y} + a_0 Y = f(x) \ \text{mit} \ f(x) \neq 0 \ \text{inhomogenous}$$

For the homogenous part we write $a_2\,\ddot{Y}_h + a_1\dot{Y}_h + a_0 Y_h = 0$ and

$$a_2\,\ddot{Y}_{inh} + a_1\dot{Y}_{inh} + a_0 Y_{inh} = f(x) \ \text{ for the inhomogenous part.}$$

With Y = Y$_h$ + Y$_{inh}$ we can now write:

$$a_2(\ddot{Y}_h + \ddot{Y}_{inh}) + a_1(\dot{Y}_h + \dot{Y}_{inh}) + a_0(Y_h + Y_{inh}) = f(x) \text{ mit } f(x) \neq 0$$

Reformulated, this results in:

$$\left(a_2\,\ddot{Y}_h + a_1\dot{Y}_h + a_0\,Y_h\right) + \left(a_2\,\ddot{Y}_{inh} + a_1\,\dot{Y}_{inh} + a_0\,Y_{inh}\right) = f(x)$$

In the first bracket above, Y$_h$ is the general solution of the homogeneous differential equation and has the value 0, for f(x) = 0.

For the second bracket, Y$_{inh}$ is an arbitrary special solution of the inhomogeneous differential equation. If we consider that the general solution of the 2nd order differential equation has two integration constants, then also Y = Y$_h$ + Y$_{inh}$ has two integration constants.
Y is thus a solution of the differential equation and has two freely selectable constants C$_1$ and C$_2$, thus Y is the general solution of the differential equation.

If one or more integration constants are given special values (constraints), a special or particulate solution of the differential equation is obtained.

Mathematics Exercise Book 14: 5.5

Excursus 1.2 Part II

For **the homogeneous linear differential equation** of 2nd order with constant coefficients, we obtain three sets of solutions due to the characteristic equation and the associated quadratic explement:

    a) The radicand is positive and provides with r_1 and r_2 two real solutions
    b) The radicand is negative and we get complex solutions r_1 und r_2 which are conjugate complex to each other
    c) The radicand is zero

to a) If r_1 und r_2 are different, we get the two different solutions Y_1 unY_2 and by forming $Y = C_1Y_1 + C_2Y_2$ we obtain the general solution. C_1 und C_2 are random real numbers

to b) The solution function is a complex function Y to the real variable x:

$Y = Y_1(x) + i\ Y_2(x)$, where the functions Y_1 und Y_2 are
are different and Y_1 is the real part and Y_2 is the imaginary part special solutions.

The general real-valued solution is then:

$Y = C_1Y_1 + C_2Y_2$. C_1 and C_2 are arbitrary constants

to c) We obtain a double root. With the exponential approach we get only one solution.

For the general solution, we have to look for a second one by varying the constants to search for a second one for example:

$$Y_2 = C_2 x\, e^{r_1 x}$$

Thus we obtain: $Y = C_1\, e^{r_1 x} + C_2 x\, e^{r_1 x}$, C_1 und C_2 are arbitrary constants

Literature

Physics

Physics Alonso/ Finn, Publisher: Inter European Edition 1977
                         Publisher: Oldenbourg, 26. Januar 2000

Mathematics

Dr. Jürgen Schlüsing Booklet 1 to 15, Publisher: Book on Demand

Mathematics for Physicists Volume 1 and 2, Publisher: Vieweg

Weltner Mathematics for Physicists
Volume 1 u. 2 Edition 2013, Publisher: Springer Spektrum

Notes